On making : 12 works by Jun Watanabe

创作谈：渡边纯作品集

中国建筑工业出版社
CHINA ARCHITECTURE & BUILDING PRESS

图书在版编目（CIP）数据

创作谈：渡边纯作品集：汉英对照 /（日）渡边纯著. 北京：中国建筑工业出版社，2013.96
 ISBN 978-7-112-15527-9

Ⅰ.①创… Ⅱ.①渡… Ⅲ.①建筑设计—作品集—日本—现代 Ⅳ.①TU206

中国版本图书馆CIP数据核字（2013）第128842号

责任编辑：武晓涛
责任校对：党 蕾 陈晶晶

创作谈：渡边纯作品集
*
中国建筑工业出版社出版、发行（北京西郊百万庄）
各地新华书店、建筑书店经销
北京京点设计公司制版
北京顺诚彩色印刷有限公司印刷
*
开本：889×1194毫米 1/20 印张：8¾ 字数：260千字
2013年8月第一版 2013年8月第一次印刷
定价：78.00元
ISBN 978-7-112-15527-9
(24071)

版权所有 翻印必究
如有印装质量问题，可寄本社退换
（邮政编码 100037）

contents / 目 录

006   On making / 创作谈

　　　　　Jun Watanabe / 渡边纯

works / 作品

014   Hiroo Flat / 広尾公寓

028   Villa Gamagori / 蒲郡别墅

040   House in Hibarigaoka / 云雀丘住宅

050   Cow Barn in Appi Resort / 安比度假酒店牛舍

060   Foreign Student Dormitory, Chubu University / 中部大学外国学生宿舍

070   Renovation of the International Congress Hall, Makuhari Messe / 幕张国际会展中心国际会议厅改造

080   Downtown Community Center, Toki City / 土岐市闹市社区中心

090   House in Shinagawa / 品川住宅

100   House in Seijo / 成城住宅

112   Taka House / 高宅

122   Villa Nakakaruizawa / 中轻井泽别墅

134   Aster Orphanage / 星形孤儿院

project summary & biography / 项目摘要与传记

148   project summary / 项目摘要

152   list of works / 作品清单

153   biography, honors and awards / 传记，荣誉与获奖

detailing / 细节

156   Villa Gamagori / 蒲郡别墅

158   House in Hibarigaoka / 云雀丘住宅

160   Foreign Student Dormitory, Chubu University / 中部大学外国学生宿舍

164   Renovation of the International Congress Hall, Makuhari Messe / 幕张国际会展中心国际会议厅改造

On making
Jun Watanabe

Since my initial architectural training, I have maintained a tendency to look at the core issue of architecture. Driven by natural disaster, the big earthquake and tsunami on March 11, 2011, the architects of Japan were fundamentally awakened to the importance of community ties ("Kizuna　絆" in Japanese). There was once a trend to regard architecture as the medium to express the architect's pedantic philosophical play.

However, after the incidents of March 11, we Japanese architects should no longer stay solely in our private realms. Instead we are now looking for architectural intensity inspired by issues in public realms. We are to serve society by intentionally getting rid of self indulgence. We should aim for the meaningfulness of architecture, which is inherent and often developed in the public realm. 'Flashy', 'eloquent', and 'talkative' are words from which my architectural pursuits surely maintain a clear distance. My works are outside the realm of self-appointed avant-gardists and inside the events of everyday life. I try to operate my work in the real world with few illusions about the messianic calling. I am interested in pursuing the depth of these architectural explorations. 'On Making' is what I set as the title of this writing. The architecture consists of materials. The material ultimately forms the architectural existence. We human beings are surely moved by great pieces of architectural work. The essential quality of architecture is, in fact, based upon the actual matter of the building. When the architecture does achieve meaningfulness, the matter of architecture inevitably speaks of some explicit themes.

It is true that small pieces in architectural element represent the consistency of the architecture. Thus, they occasionally speak about the essence of architectural achievement as Mies van der Rohe once pointed out. The making process necessitates further attention from architects. The construction always provides endless clues to enhance the philosophical exploration in architecture. It holds the sovereign quality which we architects should recognize more.

Instead of regarding on site construction as just a series of endless compromises disturbing the original design immediately before the final realization, we should intentionally face the dialectics of the immovable actuality and the pre-occupied architectural philosophy. Fundamentally important for us is the departure from the creator's arrogance. I am often intrigued by unexpected findings. These occur when I am overwhelmed by the actual presence of a building on site. To me, the key seems to be this; If we architects wish to deepen architectural philosophy, we should focus more on the dynamism of the realization process on site. The construction process should not be a minor thing that must be obeyed. Construction is more than that. On occasion, when the architect is keen enough, the reality of the building in construction brilliantly radiates. It has potential to deepen the architectural philosophy. We should try first to find the wonder itself based upon our individual curiosity. This will not come to us automatically. The more effort we make, the more enlightenment comes to our minds. Things cannot always be under our control, which is intriguing enough to be explored further. Those things which are out of our control often offer more powerful brilliance. We should intentionally leave room to appreciate those unintended developments. We can expect those happy accidents to autonomously come out in order to be recognized. We should keep our eyes open for those endlessly emerging potentials. Louis I. Kahn said "The architecture does not exist. The will of architecture exists." In 21st century society things are hastily moving on every day. In order to sense the presence of architecture one must make an effort to be silent. The certainty of the presence of architecture has grown thinner. Architects rarely care about the will behind the work, which Kahn once set as his life-long architectural pursuit. This uneasy situation creates a society floating without a sense of being anchored. We should depart from this unhealthy perspective. Ideally speaking, architecture should first emerge without any flashy arbitrariness. If we can make this happen, architecture regains a life of its own. The building inherently stands alone, occasionally in complete solitude. The building acquires a definitive condition where people are not consciously aware of its appearance. They live their daily life around and within the building itself, breathing the same air. Once the architectural work is left alone, it surely starts acquiring physical consistency with its environment.

I present these 12 works as the result of my philosophical pursuit of architecture. Here the architecture stands alone, expressing to each its own individually acquired and developed themes. Each has its own degree of expression, in powerful or subtle manners. In general, I tried to place them standing "on tiptoe" in the contemporary world of arrogance. I believe that they are indicating the presence of the ideas in built matter, hoping that they explain my argument more clearly. Overall, the real issues of tectonics and siting have become the two key driving forces carrying my architectural thinking deeper. The project Villa Nakakaruizawa (second from last), has acquired its theme through the interaction and positive appropriation with the quieter voice of the site, as well as through tectonic pursuit. Ultimately, the projects should always and essentially have their own presence. They fit smoothly and naturally on the sites as if they have found their own inherent place to be. If the finished buildings look as if they have existed there from the beginning, this is nothing but my intended goal.

受到早期建筑教育的影响，我始终保持着对建筑核心问题的关注。2011年3月11日的地震和海啸所造成的自然灾害，使得日本的建筑师对于社区的纽带（在日本称为"絆"）有了更为本质的理解。曾经有一种趋势，把建筑看作建筑师展示自己学究式的哲学思考的媒介。

然而，经过3·11的灾难，我们日本建筑师不能再单纯停留在我们自己的世界里。相反，我们需要用公共领域的议题，激活建筑的能量。我们需要有意识地停止自我沉溺，转而去服务社会。我们的目标是做有意义的建筑，存在于并且通常发展于公共事务的建筑。我的建筑追求肯定与"鲜丽的"、"动人的"和"表达丰富的"这些词汇有很远的距离。我的工作也不涉及自我标榜的前卫风格，而是融入日常生活的事物中。我希望在真实的世界中工作，而不需要救世主式的呼喊。我的兴趣是建筑领域探索中的深度。这本书的名字是"创作谈"（On Making）。建筑是由物质组成的。建筑最终存在的形式是物质的。人们总是会被伟大的建筑作品打动。建筑最根本的特质，必然是基于房屋实际的物质存在的。当建筑获得了它的意义，作为物质的建筑必然会表达出一些清晰的主题。

的确，建筑中一些很小的元素，就能表达出建筑的一致性。因而就像密斯·凡德罗（Mies vander Rohe）所说，在不经意间阐述了建筑的真髓。建造的过程需要建筑师更多的关注，它总会给建筑师提供无尽的线索，增进对建筑哲学的探究。其中蕴藏着我们建筑师需要认识的最重要的内在。

如果不再把现场建造看作经过一系列没完没了的妥协，改变原来的设计，最后只是把房子盖起来了，我们就可以直面难以改变的现实与专注的建筑哲理之间的辩证关系。对我们来说，其中最根本的原因是设计者本身的傲慢造成的背离。我总是对一些意想不到的发现着迷。而这总是发生在我被建筑工地现场的实际情况打击的时候。对于我来说，其中的关键在于：如果建筑师希望深入建筑哲理的研究，我们就必须更多地关注工地实践过程所呈现的动态。建设过程不应只是一种被动的服从，建设所具有的含义比这多得多。有时候，如果建筑师足够敏锐，房屋的建造现实就会显得熠熠发光，为建筑哲理的深入提供机会。我们需要带着求知欲，首先去发现奇迹本身。这一切不会自动降临我们身边。我们越努力地投入，思想就会得到越多的启迪。不是什么事情都能在我们的控制之中，足够复杂的事情反而能让人探寻到更多东西。不能控制的事情常常提供更为有力的闪光点。我们应该主动离开房间，去欣赏那些并非有意的变化发展。我们需要让视野更为开阔，去了解无穷的正在发生的可能性。路易·康（Louis I.kahn）说："建筑其实并不存在，存在的是建筑的意志。"21世纪的社会事物正在发生日新月异的变化。为了感受到建筑的存在，人们需要努力保持沉默。对于建筑存在感的确认正在逐渐减弱。建筑师很少考虑工作背后的意愿，而这曾经是康（Kahn）为之奋斗终生的建筑追求。令人不安的现状在社会上产生出一种漂浮不定，无所依靠的氛围。我们应该脱离这种不健康的大背景。理想化地讲，建筑首先应该摆脱浮躁而任意妄为的状态。做到这一点，建筑才能重新获得自己的生命。这样的建筑在本质上是独立的，有时甚至是完全孤立的。它能够获得一种确定的环境，人们在其中不是只关注建筑的外观。人们每天的生活既在建筑外，也在建筑内部发生着，都呼吸着同样的空气。如果独立看待建筑作品，它必然需要与环境在物质上具有一致性。

　　本书呈现的12个作品，都是我探究建筑哲理的成果。每个建筑各自独立，向人们表达它自身的需要和发展而来的理念。每个建筑都有自身的表达程度，有些是强有力的，有些是微妙的。总体而言，在充满自大和傲慢的当下社会，我希望它们都是"踮着脚走路"，悄悄地存在着。我相信，它们也都体现了建造这一事件中的观念，希望能将我的观点表达得更为清晰。总之，建构和选址的真实问题是我从事深入建筑思考中两个最主要的驱动力。倒数第二个项目轻井泽别墅（Villa Naka Karuizawa），其主题是通过建构的方式，用更轻微的声音，与基地进行互动和积极的占据。最后一点，这些项目在本质上都有其自身的表达。平静而自然地融入环境之中，就好像这是它们与生俱来该待的地方。如果它们建成后，能够看上去好像一直存在于那里的话，就是我预期的目标。

works

Hiroo Flat / 広尾公寓
2009

This multi-family residential project results from the impressive greenery of the northern adjacent property. The both exterior and interior design aim at the integration of this existing fascination with the new design. The project concerns in the issue on providing the generous relaxation for residents. North facing windows, for instance, are to view this rich greenery of the northern adjacent property. The sunken courtyard, primarily surrounded by the finely textured exposed concrete walls, is a place to enjoy the intimacy, enhanced by the overhanging ivy greenery above, which is from the northern adjacent property.

The project has considerable structural expression. The big cantilevered volume, which sticks out from the main body, is a prime example. Right above the front parking space along the street, this conspicuous structural element appears. This becomes possible by means of the floor high concrete beam within the eastern exterior wall, which runs perpendicular to the main facade. Three dimensional ceramic tiles of 5 centimeter thick are applied to the screen in the main living room. Those are structured both by the penetrating stainless steel rods and carefully applied glue. This ceramic tile screen, with moderate earth color, signifies the key ambience of the interior.

Located in the heart of the downtown residential neighborhood of Tokyo, the project contextually corresponds to the respectful townscape of the vicinity. The facade design primarily tries to achieve the understated quietness. This is to be a part of townhouses in the vicinity.

这个多户住宅项目的设计灵感，来自于其北侧相邻地块的风景优美的绿地。不论是室外还是室内设计，都试图将这一现有的美景融入新的设计之中。设计试图为住户提供更多的舒适感受。因此北侧开窗的设计都是为了能欣赏到北面的美景。下沉的庭院，由肌理精致的混凝土墙围合而成，北侧绿地生长的常青藤枝条蔓延进来，为这里增添了温馨和亲切的气氛。

设计注重了对结构的表现，其中最显著的是，从主体延伸出来的巨大悬挑构件，标志出楼前沿街道设置的停车空间。设于建筑东侧外墙内，与主立面垂直且与楼层同高的混凝土梁支撑着这一悬挑。主起居室内屏风，由5cm厚的立体陶砖拼接而成，由上下贯通的不锈钢条连接，并用胶固定。陶制屏风的淡雅色彩，形成了室内环境的主要氛围。

本项目位于东京市中心居住区的核心地带，因而对周边环境表达出充分的尊重。立面设计主要体现出低调而宁静的感受，使之融入周围的居住建筑之中。

The project explores the contrary yet intriguing coexistence of the urbanity in the front and the affluent natural forest in the back. The front volume spares the car parking space on the ground. The floor-high side beam supports this cantilevered lower portion of this main mass.

设计希望实现建筑的城市风格与其后方自然的树林之间既矛盾又相互交融的共存状态。
前方的地面停车空间由建筑悬挑的体量退让而来。建筑侧面高度与层高相同的巨梁支撑着巨大的悬挑部分。

One part side of the main facade has the 4 meters deep cantilevered volume above.

The cantilevered volume is supported by the southern side wall, which consists of a floor high structural beam.

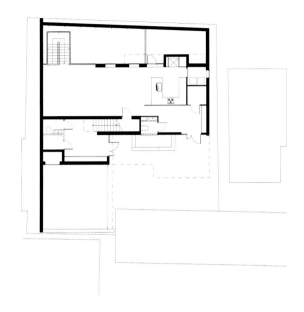

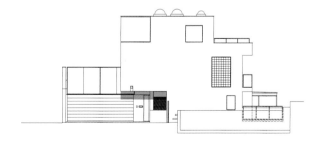
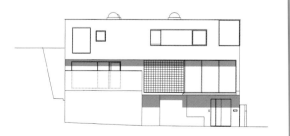

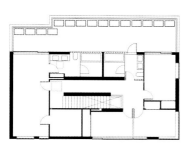
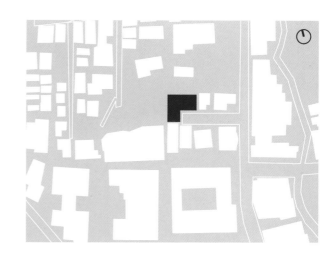
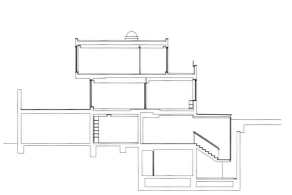
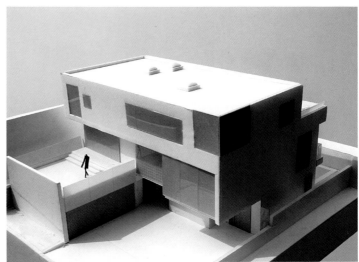

The volumetric tiles form this free standing screen. It animates the living room as well as the staircases hall.

The staircases with a sky light window above.

The staircases with three sided glass coverage appear in the intimate courtyard.

Villa Gamagori / 蒲郡别墅
2002

The design of this project is disciplined by a longitudinally running concrete beam, whose length reaches approximately to 24meters so to hold the horizontality extending flat roof and to contain the kitchen's cubic volume underneath.

The approach passage to the main entrance is sloped. At the entry point, people have two alternative choices whether to go to the right or to the left. Gradually descending on the tempered glass steps, people arrive at the living room, which is the climax of the spatial sequence. The ceiling height of this living room is approximately 4 meters. Both the front garden and the back patio are to provide the generous and relaxing ambience to this project. Three columns, to support the above described longitudinally running concrete beam, become the focus of the south facade. These structurally conspicuous elements are to animate the main entry sequence. The bridge, which runs from east to west, is noticeable in the back patio. This back patio provides the major character for the interior design. Particularly when people proceed from the entrance hall, by stepping the tempered glass steps, they are to enjoy this intimate back patio, parallel placed. After experiencing these steps, people are to reach to the living room.

This guest house, initiated by the family owned company, is to host various types of guests, including local politicians and the family's immediate relatives. The town, Gamagori is known with its coastal resort applause. In the vicinity of this project site, there is the elegant Art Deco styled Gamagori Hotel, which is dated back even to the early 1930's.

设计随着24m长的混凝土梁，沿长向展开。混凝土梁不仅承载着水平出挑的屋面，其下也遮蔽着方形体量的厨房。
通往主入口的道路是一段缓坡。在入口处有两个前进方向可供选择：向左或向右。沿钢化玻璃台逐级向下，就是起居室，也是这一空间序列的高潮。起居室的净高将近4m。前庭的花园和背面的天井都可以为室内空间带来愉悦和放松的氛围。三根立柱支撑着其上沿水平向伸展的混凝土梁，同时也成为南立面的焦点。这些以结构的方式凸显的构件，活跃了主入口的序列。连通东西的桥，是后天井的主要元素。天井的设计延续了室内设计的主要特点。人们从入口门厅拾级而下，来到与之平行布置，充满静谧、温馨的后天井，随后进入起居室。
根据客户的要求设置的客房，可以接待各种宾客，其中包括当地政要和家中来往密切的亲友。蒲郡市因其广受欢迎的海滨度假村而知名。在项目基地周边，就是建于20世纪30年代早期，具有典雅的Art Deco风格的蒲郡宾馆。

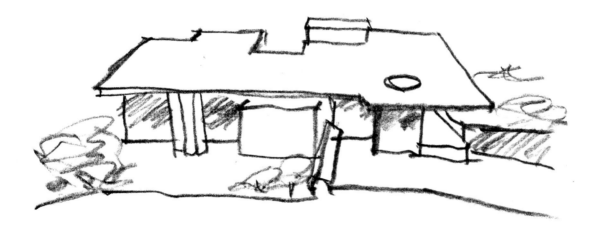

The generously extending flat roof covers the manipulated volumes and walls underneath. The longitudinally running concrete beam controls these series of manipulations, which is to generate the sequential movement and the cohesive spatial integration.

巨大的水平伸展的屋顶将多个功能体块和墙体覆盖于其下。横向延伸的混凝土梁对各种处理方式加以控制，并形成了空间上的运动，将空间紧密地结合起来。

The main garden is defined by the L-shaped plan property boundary wall.

The tempered glass staircases are between the entrance hall and the living dining room.

The tall boundary wall, made of cedar grain textured site poured concrete, provides the major character of the courtyard.

The longitudinally running beam disciplines the composition.

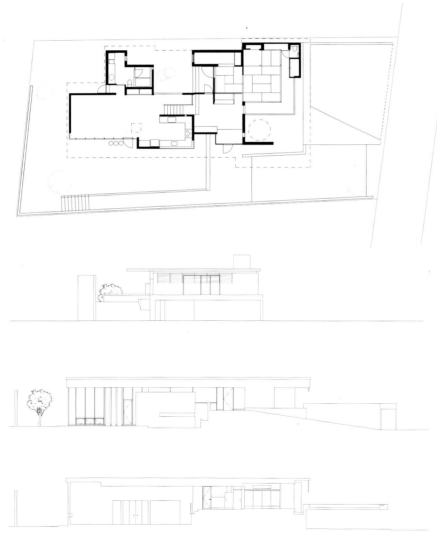

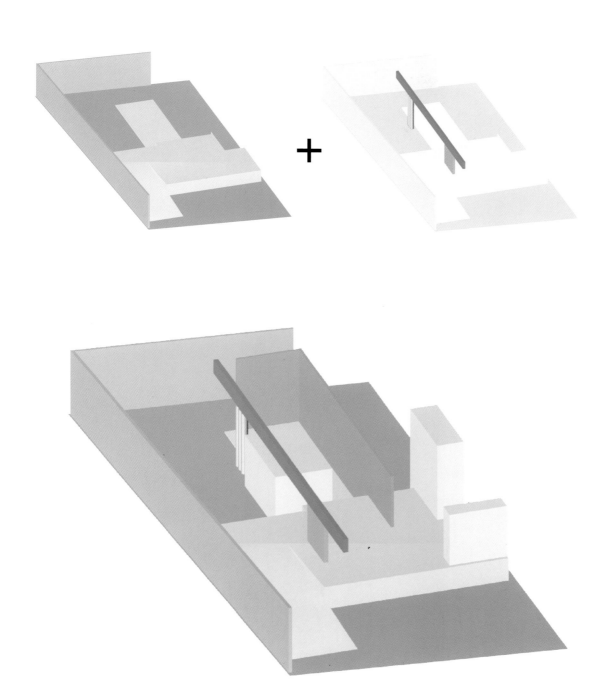

House in Hibarigaoka / 云雀丘住宅
2007

This single family house is for a middle aged couple to enjoy their relaxing living in the quiet suburban residential neighborhood in Tokyo. Since the husband has already retired, they often enjoy the world cruising. Their study rooms, individually located on the third floor, exhibit those memorial souvenirs, collected through their world cruising tours.

There are three levels in this house. The ground floor consists mainly of the living dining room and the kitchen. The second floor has bed rooms. The third floor has the big roof balcony, reached from the above described study rooms. Two intimate gardens, i.e., the front garden and the back patio, are to state a certain message clearly, enhancing the tranquil ambience of this project. The floor high inclined glass screen, placed on the upper wall of the living room, becomes the focus of the entire interior. The staircases adjacent to the back patio are the other focus. They appear right after the main entrance hall. The entry door to the Japanese traditional 'tatami'-mat room, is surfaced by the frosted mirror glass. The exterior wall is finished by the exposed concrete. Some parts have cedar grain textured surfaces. The rest has the bush hammered surface. The wood louvers, applied on the western facade with double floor height, are to control the harsh natural sun light particularly in the late afternoon of the summer time.

这座独栋住宅是为一对中年夫妇建造的，位于东京郊外安静的居住社区中，作为他们放松身心的居所。由于丈夫已经退休，他们经常去进行环球旅行。建筑的第三层只设置了书房，放置他们在环游世界的旅行当中从各地收集而来的纪念品。
这座住宅由三层组成。首层主要包括起居室和厨房。第二层是卧室。三层有一个巨大的屋顶平台，从前面提到的书房进入。建筑有两个私密的小花园，一个是门前的小庭院，一个是屋后的天井，清晰地表达出设计的目的，强化建筑静谧的氛围。设于起居室上层墙面上有一定斜度倾斜的玻璃墙体，成为室内空间的焦点。紧挨着后天井的楼梯是另一个焦点，刚好位于主门厅之后。进入日本传统的"榻榻米"——铺席子的房间的房门，采用磨砂镜面玻璃。建筑外墙是裸露的清水混凝土。部分表面采用松木纹路的肌理，其他部分为凿毛表面。西立面采用两层通高的木百叶，以遮挡强烈的，特别是在夏季午后非常炎热的太阳直射。

The project explores the issue of balancing the different four sides of the building mass. The north street-facing elevation, for instance, emphasizes the understated horizontality to maintain the continuity of the housing fabric in the vicinity.

项目探索了对建筑体量四个不同的面进行平衡的问题。例如北侧朝向街道的立面，强化低调的水平向特点，保持了与周边住宅肌理的一致性。

The site poured concrete forms the architectural mass; whereas the wood louver softens the facade design for people particularly toward the western main approach.

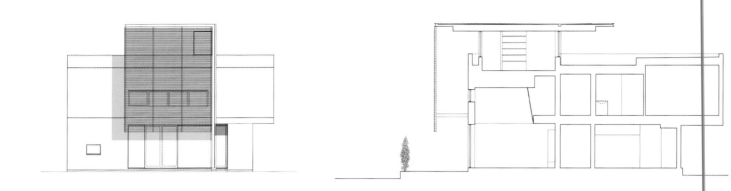
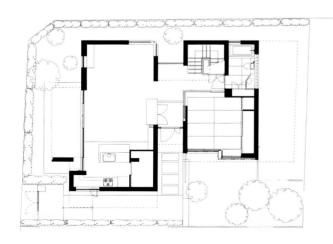
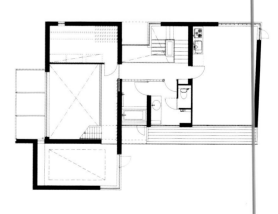

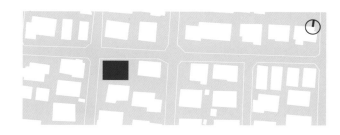

The project is characterized by
the 4-sided individual formal
statement.

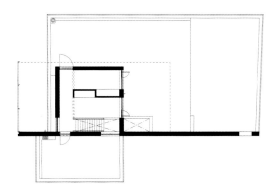

The living dining room has the double height.

The staircases surround the lighted displaying base to accept the sculptural object.

The ground floor has the traditional 'tatami' mat room with the entry sliding door, which is finished with a frost mirror glass.

Cow Barn in Appi Resort / 安比度假酒店牛舍

2000

This cow barn serves both for milk producing cows themselves and for summer tourists in Appi resort. Cows become more productive, moving in this newly constructed barn. They have been refreshed. For the sake of the cow farming enterprise, the milk quantity increase is important particularly for the decision toward constructing this new barn. However, the key intention lies on the further expectation for summer tourists. The project wishes to affect them. Based upon the market analysis, summer tourists in general prefer the straightforwardness of the cow farming enterprise. They may dislike the attraction-oriented flattering. Thus, this new cow barn should seriously stay as a milk producing factory. Those summer tourists should look at the real processing of the cow farming. The project, with a robot operated milking machine, houses 44 cows. This machine, made in Sweden, was the third first introduction to Japan.

Situated in the highland's landscape, the configuration of the roof responds to the site context. On the people's approach from the hotels, the roof configuration deliberately captures the undulation of the meadow slope in front. It is gradually raised to reach to the climax where the series of ventilation slit openings are placed. This area has overhang to cover the visitor's passage along the southern edge of the cow barn. Visitors are allowed to look at the cow farming activity only from this passage, covered by the overhang. They are to be controlled to keep staying outside. Limited numbered visitors, guided by the responsible staff, are specially allowed nearly to see the robot machine which is completely automated for milking cows constantly for 24 hours.

该养牛场既用于饲养奶牛，同时也可供夏季来安比度假酒店牛舍的游客参观。奶牛迁入这座新的牛舍后精神更好，产奶量也有所增加。对于养牛场来说，增加产奶量是决定新牛舍建造的重要指标。然而，设计最主要的意向还是为了吸引夏季前来参观的游客。根据市场分析，夏季游客对养牛场淳朴的风貌通常很感兴趣，反而不喜欢那种一看就在招揽生意的形式。因此，新建的牛舍需要保持其工厂的特点，让夏季来的游客看到奶牛场的真实生产过程。该牛舍可容纳44头奶牛，并安装有自动挤奶机械设备。设备产自瑞典，是第三台新引进到日本的产品。

牛舍建于安比高原的自然风景中，屋顶与基地的周边环境相呼应。从宾馆前往牛舍的道路眺望，屋顶的形态仿佛在模拟近旁起伏的草坡，逐渐升高，在顶部设有一系列通风口。屋顶的出挑部分可以遮蔽牛舍南侧的游客通道。使游人沿通道参观养牛场的各种活动，参观范围仅限于室外。少量游客在工作人员的带领下，经过允许，可以靠近24小时工作的为奶牛挤奶的全自动设备。

The project explores the roof form to achieve the formal continuation of the sloped meadow in front. The opposite side of the roof provides the ventilation slit opening. This part also provides the overhang adequately to cover the tourists` passage underneath.

该项目的屋顶形式延续了屋前草坡的起伏。屋顶的另一侧开设了通风口。该部分向外悬挑，遮蔽其下方的游客通道。

The project explores the roof form.

The right image shows the ventilation slit which runs horizontally on the upper portion of the side bent. There is a tourists` passage on the bottom of this triangulated protrusion.

The natural light comes from the horizontally running window and from the round shaped sky light windows above.

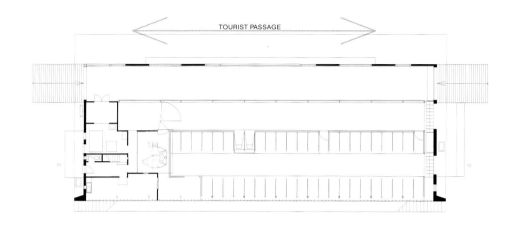

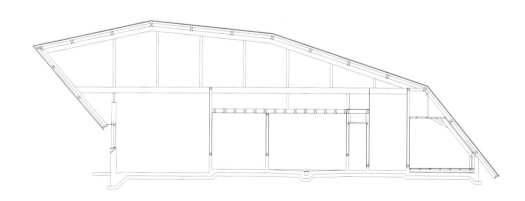

The wood truss beams are applied.

The steel structured framing holds one side of these wood truss beams.

Foreign Student Dormitory, Chubu University / 中部大学外国学生宿舍

2003

This international student dormitory houses 29 residents. The pavilion like appearance results from the centrally located hilltop condition in this dormitory and staff housing area of Chubu University campus. The building is for those international students, who often stay in this dormitory in the campus for their entire four years educational period. Comfortably living together with other students, they often are reluctant to move out to start their new living outside of the campus.

The project considers about the special circumstance of these international students. Spiritually, the architecture should become a reliable symbol for them. Despite their hardship, i.e., the loneliness of being abroad, those international students make their efforts to accomplish their study. Stacked in three levels along the southern edge of the complex, 29 private rooms were aligned in a boomerang-shaped plan which implied the palm of a hand being cupped around something precious. The three-story high main space (the dining and lounge area) cradled by the 'palm' has a large glazed surface that acts as an 'interface' providing both a connection with and separation from the natural wooded area outside. The northern block which houses the first-floor kitchen and the second and third floor shower rooms heightens the effect of the light pouring in from the skylights above. The bridge on the second floor that crosses under the skylights serves as a focus of the interior space of this project.

国际学生宿舍可容纳29名住宿学生。这座宿舍和中部大学校园的员工居住区都位于山顶的中央，考虑到环境特点，大厅的外观与之呼应。这座建筑是为那些四年学习生活中大部分时间都会在校园中、宿舍中度过的国际学生建的。由于在这里便于和其他学生共同生活，他们一般不愿意搬到学校外面去。

项目考虑到国际学生的特殊情况，建筑要成为他们精神上可依靠的象征。国际学生会尽力忘却身在异乡的孤独等各种烦恼，把精力用在完成学业上。建筑南侧三层高的部分容纳了29个住宿单间。宿舍的排列呈飞去来器的形状，平面如同一只手，呵护着内部空间。"手掌"环抱的主空间（餐厅和休息区）为三层通高，大面积的玻璃窗既是室内外的分隔，又将外部自然的林地风景引入室内。建筑北侧部分的一层为厨房，二、三层为淋浴室，大厅的屋顶天窗为这一侧提供了良好的自然采光。二层的廊桥位于天窗正下方，也成为建筑室内空间的视觉焦点。

The hilltop location of this building signifies the focus in the Chubu University Campus. The building terminates two axes; one for the class room buildings zone and the other for the dormitory and the staff housings zone.

由于地处山顶，建筑成为千叶大学校园的视觉焦点，同时也作为教学区和教师居住及其他宿舍区两条校园轴线交会的终点。

The natural light introduced by the high sash window animates this double height main dining room.

This sash faces to the cliff, covered by the dense natural forest.
(the following page)

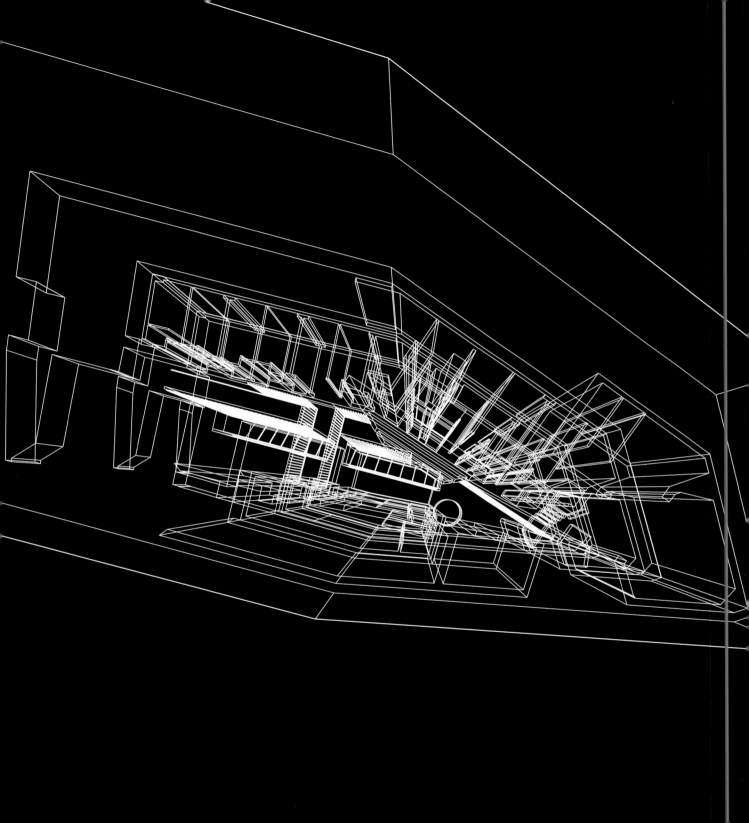

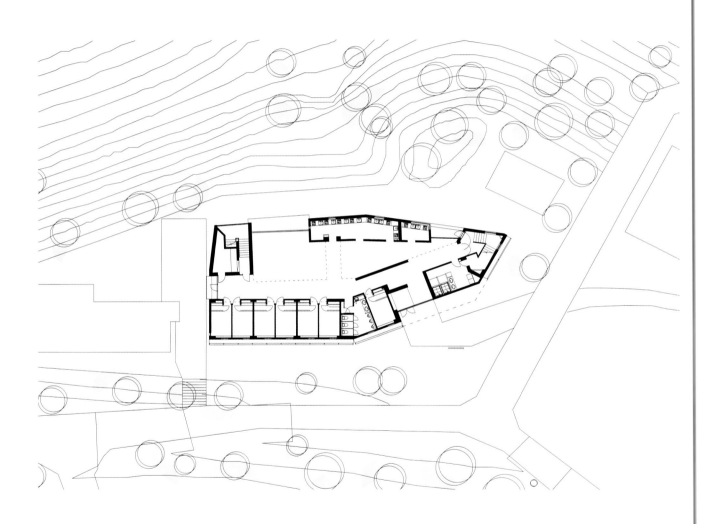

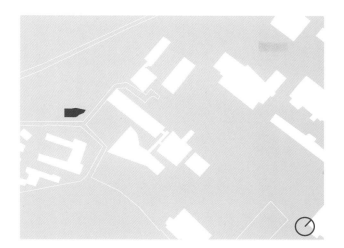

The northern windows face to the cliff.

The southern main facade faces to the surrounding campus buildings, consisting mostly of classrooms.

Renovation of the International Congress Hall, Makuhari Messe / 幕张国际会展中心国际会议厅改造

2011

This renovation project resulted from the urban maturity of the Makuhari CBD, which is regarded as the sub-center of the capital, Tokyo. The project spans from the issue of interior design to that of the urban design. The Japan's largest convention center, MakuhariMesse originally contained the various restaurant facilities, which even hosted V.I.P guests. This situation was inevitable, because the surrounding properties were completely unoccupied in the first couple of years. In the beginning, there was no facility, which enabled to support convention activities. Time has passed and subsequently a couple of commercial developments occurred. The vicinity has finally been the matured urbanized area. By having significant number of tall office buildings, the area is now regarded as the sub-centerof the capital, i.e., Tokyo.

Thus, the project replaces the existing main restaurant with the newly installed 6 meeting rooms. It aims for the updated and animating facade, particularly to the main approach from the transportation hub, i.e., the train station, as well as from the hotel accommodation area. The interior of the project has the design focus of the inverted corn-shaped column, located in the lounge area. There are two other noteworthy design explorations. One is the design of the gate from the existing part. Passing this threshold, people prepare themselves toward the convention activity. The gate itself is made of the shiny black stainless steel. Second is the design of powder rooms, which is to present the revised newness of this 20 year-old big convention center.

这个改造项目的需求是随着幕张市CBD的城市空间的逐渐成熟而产生的，幕张市是日本首都东京的副中心。设计从室内设计一直扩展到城市设计。作为日本最大的会议中心，幕张国际会展中心在建成之初，可以容纳各种餐饮设施，甚至有专门场所接待VIP宾客。这种情况是因为，在建成初期的若干年内，周边用地还完全没有得到开发。而在最初的设计中并没有辅助会议活动的设施。随着时间的推移，周围逐渐开展了商业设施的建设，最终形成了成熟的市区。会展中心周围林立的办公建筑，使该地区成为首都东京名副其实的副中心。

因此，该项目是用6个新建的会议室取代原有的餐厅。其目的是更新外立面，使建筑焕然一新，尤其是对面向从交通枢纽，即火车站以及酒店居住区前往会展中心的主要道路的正立面。项目室内设计的焦点，是设于门廊的倒玉米形的圆柱。此外还有两点值得注意的建筑尝试。一个是从现有部分前往新建区的大门设计。门由光面黑色不锈钢制成，穿过这个大门就在提示人们开始会议活动的准备。设计的另一个特点是盥洗室，为这个已有20年历史的会议中心带来改造后的新气象。

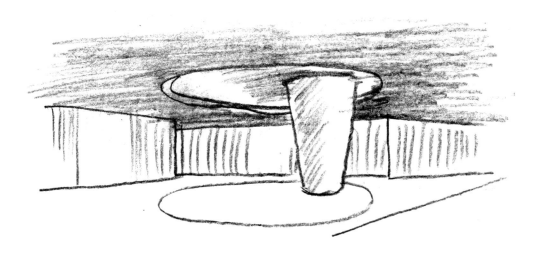

Because of its prominent location in this big convention facility, MakuhariMesse, this interior architectural project even affects the urbanity in the vicinity. Particularly, the inverted corn shaped column works as the marking element to be referred by people, who is even at the sidewalk of the frontal main street.

由于在幕张国际会展中心，这一大型会议设施中居于显著位置，该室内建筑设计项目甚至影响到了周边的城市特征。尤其倒玉米形的柱子，由于行人从门前的大道上一眼就能看到，成为大家争相参考模仿的标志性元素。

The planning to follow the requirement results in leaving this one isolated column.

The project focuses to design this column to be the symbolic element in the lobby space.

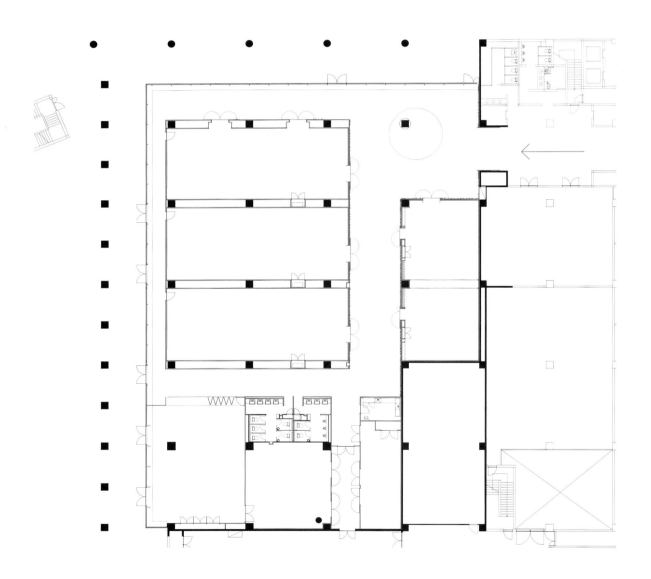

6 meeting rooms are planned, with the generously allocated lobby spaces.

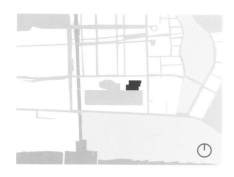

The louver is ended by the vertical frame made of the shinny polished metal. There are a couple of indirect lighting niches.

The design motif of louver wall is adopted for the main corridor.

The niche for the indirect light is finished by the shiny stainless steel panel.

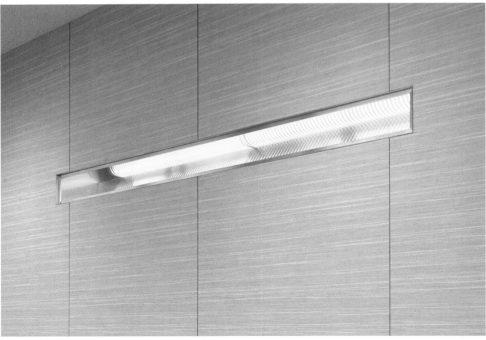

Downtown Community Center, Toki City / 土岐市闹市社区中心
2003

The project aims at the revitalization of the downtown, Toki city in Gifu prefecture, which was once prosperous with its well-known ceramics industry. It is a downtown community center, with an essentially important covered market place attached. The building's main volume consists of meeting rooms, shops, a gallery, and a café. Having class rooms in the heart of the city, the city council wishes to enhance the educational opportunity, particularly for retired senior citizens and young mothers. The program even tries to foster the kid-raising, expecting to improve the problem of the aging society, which typifies this downtown area of Toki city. The attached 400m^2 covered roof structure, made of glasses, is to provide the open air gathering space underneath. Outdoor market activities are held in the every weekend. Accepting many citizens at this particular occasion, the area becomes an important urban focus.

Although the weak commerce has caused the deterioration of this area in the past two decades, the valuable structural clarity of the urbanity always exists in this downtown area. Noteworthy is the natural resources like a beautiful river. Sandwiched by this Toki River and the railway with Toki Station on it, this downtown area has potential to receive the zeal toward further urban redevelopments. It may easily accept new condominium developments, which does produce desirable profit. Thus, for the purpose toward the downtown revitalizing, the new construction of this community center is eagerly awaited, as the initial anchoring project. Toki city ought to decide the actual start of the construction of this new downtown community center, with no further hesitation.

本项目的目标是复兴岐阜县土岐市的市中心，这里曾因知名的陶瓷产业而繁荣一时。项目是一处闹市区的社区中心，并且需要建造一个重要的大棚式市场。建筑主体容纳会议室、商店、展廊、咖啡厅等功能。市议会希望通过在市中心设置教室，增加教育机会，主要针对退休的老年人和年轻的母亲。这一计划希望通过促进育儿，改善老年社会的问题，这是土岐市中心地区的典型问题。通透的玻璃屋面底下提供一处面积400m^{2}的开放空间。每个周末这里都会有露天集市活动。很多市民会来到这里，使之成为重要的城市焦点。

尽管在近20年间，商业萎缩导致该地区的衰败，这一有价值的建筑还是能够体现市中心始终存在着的都市氛围。自然资源，即旁边美丽的土岐河，是其中值得关注的要素。土岐市的中心区夹在土岐河和土岐火车站的轨道之间，有潜力随着城市的进一步发展获得更多的活力，便于接受新的共同发展方式，获得令人满意的收益。该项目的缘起，就是为了实现市中心复兴的目标，对新的社区中心的迫切需求。土岐市应该减少迟疑，尽快开始新的闹市区社区中心的建设。

The project explores the integration of the main volume with the generously opening big roof portion, which is covered with glasses and is to become the important weekend open market place to accept many local citizens.

项目试图将主要体量和非常开敞的大屋顶相结合。玻璃屋顶所覆盖的场地,是当地市民重要的周末露天集市所在地。

The street facing main facade has a colonnade space.

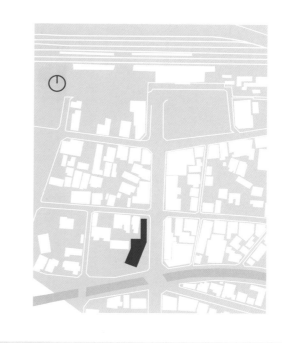

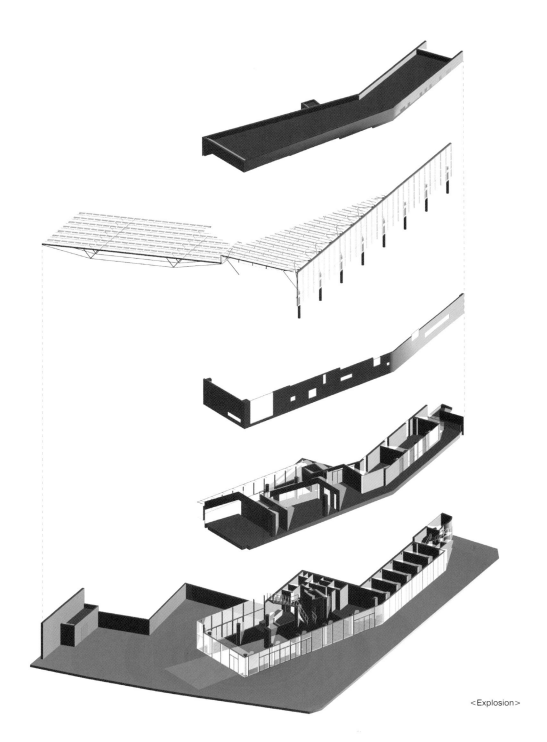

<Explosion>

The glass covered market place facilitates the community life in the vicinity.

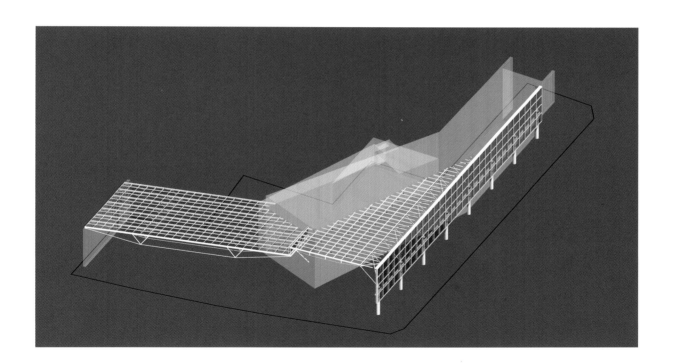

The roof difference provides an additional truss beam.

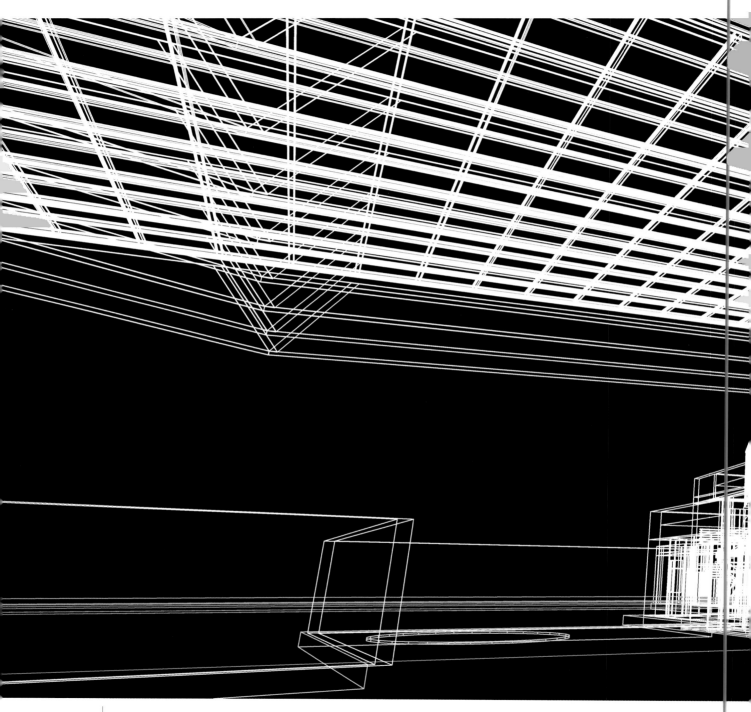

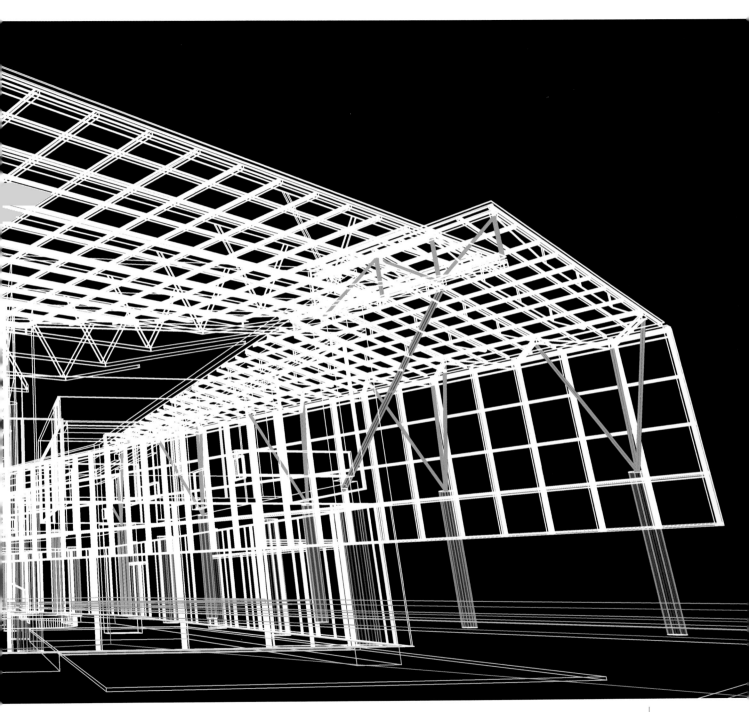

House in Shinagawa / 品川住宅
2011

The design of this steel structured single family house, located in the southern outskirt of the downtown Tokyo, focuses on the eastern facade, facing to the front street. The layering effect is intended. The applied 3 layers are to provoke their own individual expressions.

The first frontal one, which is the see-through screen, is made of the fiber glass grating. It controls the visibility, yet allows the natural ventilation. The second layer is the first front wall of the building. It is finished by the lightly colored stucco to present the basic coloring direction of the project. The third layer is the recessed wall of the building. The steel exposed outdoor staircases, which start from the second floor balcony, reach to the roof balcony, where the panorama view opens. This exterior staircases are sandwiched by the above explained the first and the second layers. This second floor located and a half balcony importantly becomes the main gathering spot. The big balcony, placed between the second and third layers, and attached to the L-shaped living and dining room, animates the entire house, becoming the focus. There is a wide open and generous entrance porch, which works as the additional car parking for guests.

The spatial organization of this house aims for the interactive and experiential quality. This should contribute to the clients' wish toward the 'family tie'. By having this fundamental, the house become a gathering spot. The clients wish to enjoy themselves, surrounded by their close friends and immediate relatives.

该项目位于东京市中心以南的市郊地区，是一处钢结构的独栋住宅。设计的重点是其朝向前方街道的东立面，通过使用三层构造的叠加，形成建筑独特的外观，营造出层层叠叠的效果。

这一层叠立面的第一道幕，是由玻璃纤维制成的通透的格栅。格栅降低了内部的可见度，又能保持自然通风。第二道幕是建筑的第一道外墙。墙体的浅色涂料，也确定了整座建筑的基本色彩倾向。第三道幕是建筑内退后的墙体部分。二层平台和屋顶平台之间以充分暴露钢结构的室外楼梯相连。人们可以在屋顶上俯瞰周围的全景。室外楼梯夹在前面提到的立面的第一道和第二道幕之间。位于二层的半室内化的平台因而成为建筑主要的聚会空间。这个大平台处于第二道和第三道幕之间，与L形平面的起居室和餐厅相邻，活跃了整座建筑，成为其中的焦点。入口门廊相当开放、通透，可作为来访宾客的停车位。

建筑的空间组织意在产生相互交流和空间体验的特质。以满足业主希望住宅成为"家庭纽带"的目标。根据这一原则，住宅成为一处聚会的场所。业主希望能在这里和他们的亲朋好友一起，尽情享受生活。

Architecturally, the frontal screen, made of perforated metal, is intended to become the membrane to assist the residents properly to be connected with the urbanity of the immediate surroundings.

由通透的金属格栅组成的外幕墙,就像建筑的一层膜,帮助住户与周围临近的都市空间建立了恰当的联系。

The key design of this project is the screen made of fiber glass. It covers the main balcony on the 2nd floor and holds the steel staircases reaching to the rooftop.

The layering effect is intentionally generated by the frontal screen, made of fiber glass grating.

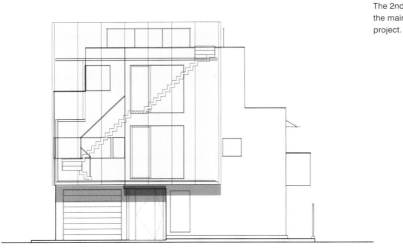

The 2nd floor balcony becomes the main gathering spot in this project.

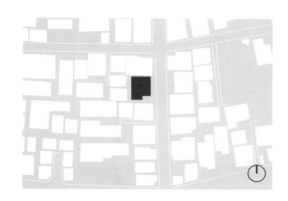

The sky light window behind the sofa animates the living room.

House in Seijo / 成城住宅

2009

The project, located in the residential neighborhood in Tokyo, is for a young couple with pets such as a cockatoo, a cat and tropical fishes. The design is resulted from the compositional development of the volume of the second floor, which primarily houses private bedrooms. This part is intended to generate the rhythmical movement responding to the contextual issues. The project tries to enhance the interactive mitigation with the surrounding streetscape. Most streets there are essentially tree lined. The project tries to provide the relaxing ambience, following the clients´ wish.

On the south east corner of the property, there located a city registered camphor tree. This corner symbol tree is to initiate the architectural promenading. The individual scenes, whose episodes are deliberately coordinated, are to appear sequentially along the expected promenading route. The patio-facing staircases are located in the central portion of the house. The patio space is intended to provide the tranquility. This is the space to enjoy the intimacy. On the contrary, the airy living room, with a kitchen behind the central tall cabinet , occupies the major part of the ground floor. This central tall cabinet, sit in the middle of the void space, anchors the house itself. There is the top-located study room, from which the roof balcony can be accessed. Both the roof balcony and the ground floor garden are intended to animate the house with their affluent greenery. The western exposed second floor balcony is to serve as a major intermediary space, where people enjoy the sunset drinks, viewing the farm land of the western adjacent property.

本项目坐落于东京的一处居住区内，是为一对年轻夫妇设计的，他们热爱饲养宠物，同时拥有一只风头鹦鹉、一只猫和许多热带鱼。整座建筑的设计，是由二层体量的组合、发展形成的结果。二层的主要功能是私密性的卧室，在设计中通过有韵律的动感变化，回应周边环境的脉络。项目尝试与街道景观之间进行更深入的交互、融合。周边大多数的街道沿街都遍布绿树，景色宜人，设计的目标就是为这对业主夫妇提供放松的氛围。

用地的东南角，有一棵已由市政府登记的樟脑树。作为街道转角的标志物，这棵树成为建筑空间序列的起点。在这条经过设计的沿建筑行进的路线中，空间序列的每一幕场景，都经过了谨慎的协调。建筑的中心部分，是朝向天井的楼梯。天井空间为业主带来宁静、享受私密感。而另一方面，作为首层平面主体部分的起居室，则呈现出开放的状态。位于平面中心的通高的柜子，分隔出其后方的厨房空间。居中的高柜位于吹拔空间内，控制着整个室内空间。顶层设有一间书房，可以到达屋顶平台。不论是屋顶的平台还是首层的花园，都通过其中多姿多彩的绿化，活跃了建筑的氛围。二层西侧开放的阳台是一处主要的室内外过渡空间，业主可以在这里品酒休憩、享受落日美景，也可俯瞰毗邻用地西侧的农地景致。

The volume of the second floor, which is to be the major formal expression, is raised. It appears as if it floats from the ground.

抬高的二层体量,是形式表现的主要部分,仿佛漂浮于底层之上。

The main garden is defined by the L-shaped plan property boundary wall.

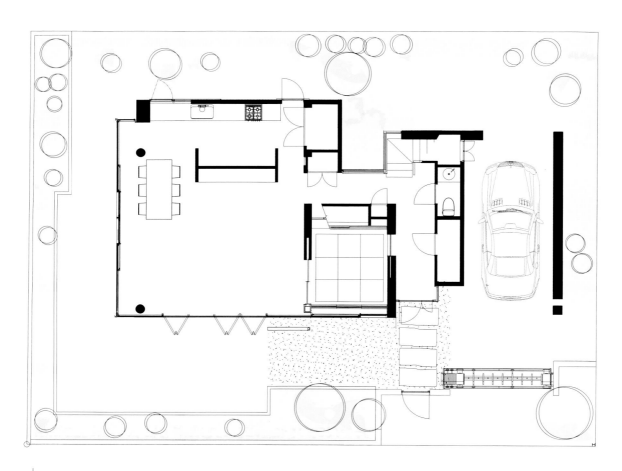

The interior view of the traditional 'tatami' mat finished room.

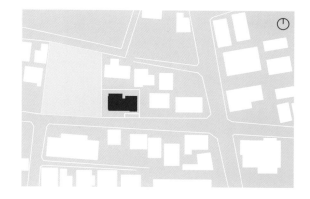

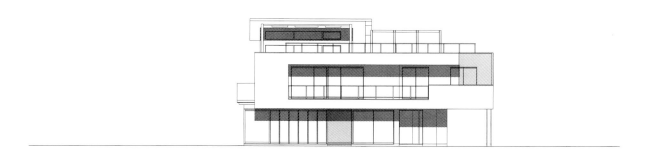

The staircases face to the intimate patio.

The glasses particularly shine in the evening.

Taka House / 高宅
2007

This is a conversion of a 30 years old single family house. The project focuses on the tuning of the interior spaces, suitable for the clients, who are a couple after the kids-raising period. The husband and his wife have initially expressed their wish to remodel their living house at that time, which looked ordinary even for them. The middle aged husband is a healthy and active man. He frequently enjoys the sport car racing by himself. The Suzuka circuit, which is a well-known circuit place, is the place where he goes almost in the every weekend. His wife, on the contrary, is a modest lady who likes to enjoy her own quiet living in the private realm. Thus, the couple has started looking after the quality oriented living place, through altering their old house.

First, they have specified the floor finish material of the main space. The floor is to be finished by the tiles and stones. This essentially initiates the project. It has affected the rest of the design decision. They express their dislike of the wood grain appearance on the any surface of the interior finish. Under this basic, the project tries to harmonize the things, simply and cleanly. The intended ambience is that of a cozy residential hotel in the heart of the large cities. Regarding to the program, they particularly wish to enlarge the kitchen and the bath room area. The kitchen should be an open kitchen with the island type counter in the middle. There is a trial to make the kitchen counter to be a sculptural object in the generously laid out one room, which consists of living and dining spaces. Architecturally, the double layered glass screen with the LED lined lighting fixture on the top becomes the focal element in the living and dining room.

本项目是一座有三十年历史的独栋住宅的改造设计。设计的要点是协调其室内空间，以满足业主夫妇在养育孩子阶段内的需要。业主夫妇最初表示，他们希望改造起居空间，因为这里对他们来说显得太普通而乏味了。男主人是个健康活跃的中年人，爱好赛车运动，几乎每个周末都要到著名的赛车场所——铃鹿赛道去。相反，他的夫人是一位非常温和的女性，喜欢在私密的空间内独自享受宁静的时光。因此，这对夫妇最初的目标就是能够通过改造他们的老宅，形成一处有品质的居住场所。

他们在设计初始就为主要空间指定了饰面材料，希望以瓷砖和石材铺地。这成为整个项目的出发点，并对建筑其他部分的决策产生影响。业主夫妇强调他们不喜欢木材的肌理，因而不希望室内的任何表面是木质饰面。根据这一前提，设计尝试通过简单而清晰的方法协调各个部分，营造出一种大城市中心地带舒适宜居的宾馆的氛围。对于建筑布局，他们还特别要求加大厨房和浴室的面积，厨房应采用岛式布局，在厨房的中心设置橱柜。由起居室和餐厅组成的大空间在整体上采用了平铺直叙的设计手法，设计尝试让这个橱柜成为其中一个雕塑化的部件。在建筑上，上部排列着LED照明设备的双层玻璃幕墙，成为这一起居和餐厅空间的视觉焦点。

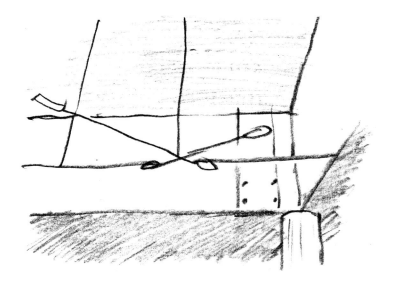

The project explores both the reflection and the translucency of the glasses. The glasses are used primarily at the free standing screens and at the staircase steps. The wall, finished by the shiny polished Italian stucco, which includes the marble powder ingredient, synchronizes the above described effect of the glasses.

项目探索了玻璃的反射和透射特性。玻璃主要是用于独立的隔断和楼梯梯步。墙体粉刷采用光滑的含有大理石粉成分的意大利涂料,具有和前面提到的玻璃同样的特性。

The project explores the effect of the double glazed glass screen with the LED lighting fixture in between.

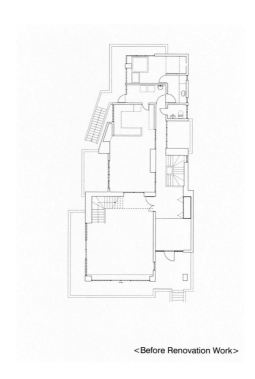

<Before Renovation Work>

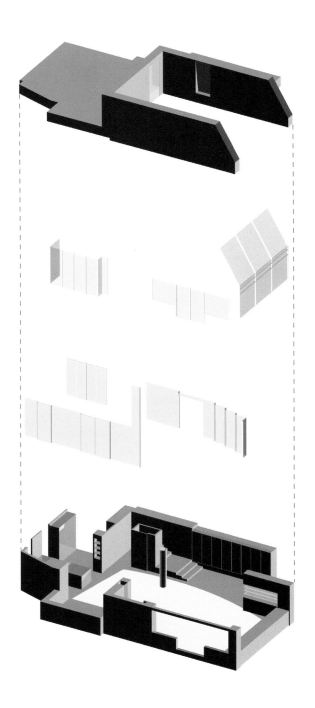

The double height living room has the custom made architectural lighting.

The pre-existing situation before the renovation work starts.

The renovated situation after the project completion with furniture in it.

The living room without furniture.

The staircases with the effect of blue color artificial lighting.

The translucent glass screens at the entrance hall.

Villa Nakakaruizawa / 中轻井泽别墅
2004

This project is a single family house located in Karuizawa, a well-known highland resort town with approximately one hour rapid train ride from Tokyo. The house is for the clients who wish to relax and enjoy their country living, surrounded by the pre-existed beautiful natural forest. When the project was initially started, it was a mere weekend house. But gradually, the clients wished their main house to live in continuously. There is the rich natural forest in the northern adjacent property. This, even with a natural stream in it, has significantly influenced the design. The lower located north facing picture window, along the floor level of the main entrance, provides the vista toward this northern located natural forest. The clients could enjoy the serenity, which is particularly intended in this project.

Nestled in the bottom of a small hill along the western property, the L-shaped plan is adopted. It holds the stone finished front terrace, from which the architectural promenading starts. The main entry, located in the middle portion of this L-shaped plan project, is to allow the spatial flow reaching to the climax of the interior, i.e., the living room with the doubled height ceiling. It becomes the focal point of this house. There is a gallery running 2.3 meter above the floor level of the living room. On the right corner of the fireplace in the living room, few steps are placed in order for people to reach the lower located guest rooms.

Regarding to the facade design, there is a continuous high side window with horizontally running aluminum louvers. This is to provide the natural light in the ceilings of kitchen and entry hall.

该独户住宅坐落于中轻井泽，这里是著名的高原度假小镇，距离东京约有一小时快速列车车程。客户希望在森林环抱的美景之中得到放松，享受乡村生活。项目刚开始的时候，客户只是想在周末到这里短暂居住。但随着项目的进展，他们希望能够把这里作为主要住宅，长时间地居住。别墅北面的森林提供了丰富的自然景色，一条小溪流淌而过。这些都显著影响了建筑的方案设计。朝向北侧的景窗，沿主入口的地面高度开设，将北面森林的全景尽收眼底。让客户享受静谧的时光，成为设计的主要目的。

建筑位于一座小山脚下，沿用地西侧展开。布局为L形。前面的平台由石材覆盖，建筑各部分散布其上。主入口位于L形平面的中部，使空间流线在室内形成高潮，起居室净高为两层，是整座别墅的中心，周围环绕标高为2.3m的走廊。起居室壁炉的右侧角部设有几步台阶，将人引向较低处的客房。

根据立面设计需要开设的水平向连续侧窗，其外侧设有铝质格栅，能够将自然光反射到厨房和入口门厅的顶棚上。

The project tries to develop the cutting section. It implies the ancient Greek's 'megalon' complex, which holds the secure part for people on the back to allow the desirable vista toward the front.

设计由剖面发展而来,暗示古希腊"梅加隆"式复合形式,将人们生活的私密部分放在后面,而将需要欣赏的主要景观放到前面。

The house is inserted in the existing rich greenery.

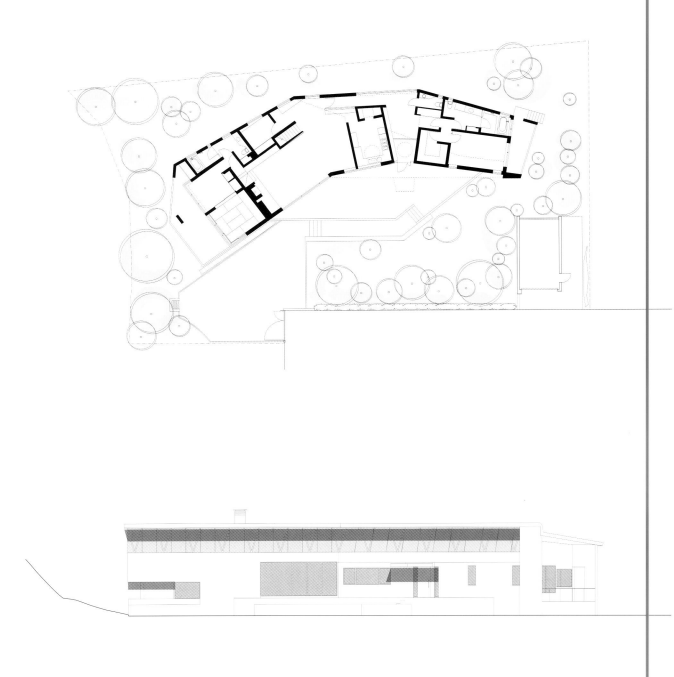

The configuration of the plan suggests as if it holds something precious on the palm.

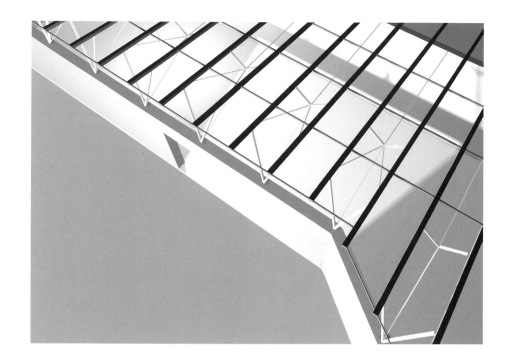

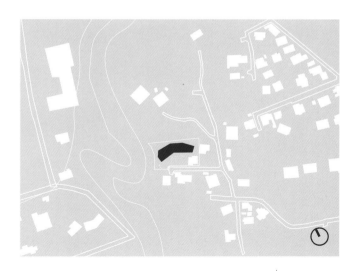

The horizontally running high side window is the conspicuous element of the main facade.

The exposed site-poured concrete, finished with bush hummer surface, is applied even to some interior walls and ceilings.

The beam string structure participates in the ceiling design of the living room.

Aster Orphanage / 星形孤儿院
2005

This small-sized orphanage has 3 design points. First, the understated presence of the building in the neighborhood is explored. The project wishes to acquire the appearance of a single family residence, blended in the regular residential fabric in the vicinity. Leaving the previous living style in the old-styled camp, contemporary orphans are to be normalized. They should naturally live in a small sized 'house', which almost looks like a single family home. Secondly, the project explores to make this new orphanage open itself to the surrounding, in order to be welcomed by the community. The project develops the devices to facilitate the positive warmth from the surrounding local community. An example is the wide opened intermediary space, which is right on the lower portion of the main facade. It faces to the intersection of the local streets. This open space, with a tall ash tree, is intended to be a bazaar/ party space to accept street walking local inhabitants. Thirdly, the project allocates a certain portion for those who are in the transitional period. Some families wish to reconstruct their family tie. However, because of the spatial limitation, the care for those has been left over. The project is intentionally allocating a certain portion for those families, which has its own independent entry, facing to the southern garden. Both of those mother and her child/ children necessitate the warming up period comfortably spending under the umbrella provided by the counseling experts. The orphanage tries to follow this essentially important programmatic requirement, as well.

这个小型孤儿院的设计有三个要点。第一，建筑与环境相融的简朴外观，是经过深思熟虑的结果。项目试图融入周围普通住宅的肌理。因为当代孤儿院的设计，已经从传统营房式的居住方式，向更为正常的方向发展了。他们应该生活在那些小型"住宅"中，看上去就像一般的住家。第二，新孤儿院的设计更为注重向周边环境开放，希望能被社区所接纳。孤儿院各部分的布置都考虑到增进与周围居民的情感联络。例如，在主立面较矮的部分，设置了开阔的过渡空间，面向街道十字路口。开敞的空间加上高大的白蜡树，形成了一处可用做聚会的集市，吸引街道上往来的居民。第三，孤儿院为过渡时期的家庭提供了一部分住所，有些家庭希望在这里建立他们的家庭纽带。但以往因为住所有限，这一意愿很难得到满足。这个孤儿院则为这些家庭设置了专门的区域，并设有独立的入口，朝向南侧花园。不论母亲还是孩子，在专家的咨询指导下完成预热期的过渡是非常有必要的。孤儿院遵循这一重要的基本需求，设置了此类场所。

The corner facing porch with the anchoring symbol-column is the main architectural intervention following the programmatic requirement.

朝向门廊的一角,门廊内带有标志性的柱子反映了建筑追随功能需求的设计。

The wooden louver covers the second floor balcony.

The wooden louver experientially affects the residents´ balcony activities.

The penetrating ash tree becomes the symbol. It works importantly for the exterior design, especially at the north east corner of the property.

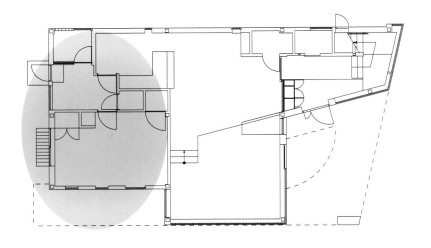

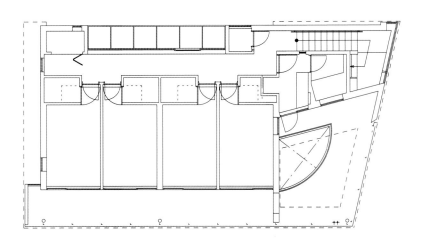

The pink colored zone on the left indicates the specially allocated portion of this project. It is for those uneasy families who are in the transitional period.

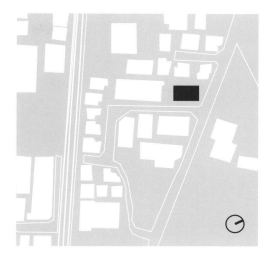

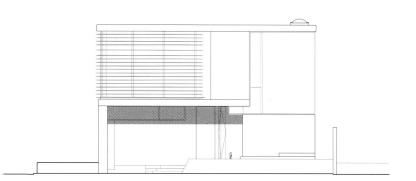

The sky light window above animates the kitchen area.

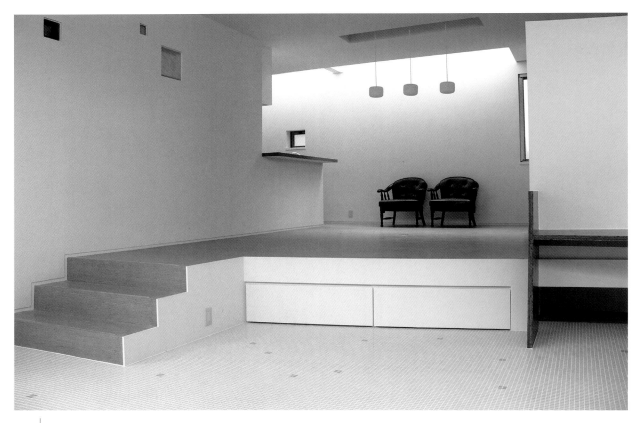

There is a level change. The living room is 3 steps lower than the dining room.

project summary & biography

Multi Family Housing		Shibuya, Tokyo, Japan
2009		
	Hiroo Flat	Site Area : 318.64m^2
Good Design Award 2011		Total Floor Area : 552.20m^2
2010 Selected Work(Design Award), Japan Institute of Architects		

Guest house		Gamagori, Aichi, Japan
2002	**Villa Gamagori**	
2003 Gold Medal, Bienal Miami+Beach Residential Category		Site Area : 230.94m^2 Total Floor Area : 580.14m^2

Single Family House		Higashikurume, Tokyo, Japan
2007	**House in Hibarigaoka**	
		Site Area : 245.94m^2 Total Floor Area : 194.95m^2

Cow Barn 2000 2002 Selected Work(Design Award), Architectural Institute of Japan	**Cow Barn in Appi Resort**	Appi, Iwate, Japan Site Area : 42,550.60m² Total Floor Area : 759.73m²
Student Dormitory 2003	**Foreign Student Dormitory, Chubu University**	Kasugai, Aichi, Japan Site Area : 7,287.73m² Total Floor Area : 1,019.84m²
Conference Hall 2011	**Renovation of International Congress Hall, Makuhari Messe**	Makuhari, Chiba, Japan Site Area : 173,191.47m² Total Floor Area : 1,656.30m²

Community Center		Toki, Gifu, Japan
2003		
	Downtown Community Center, Toki City	Site Area : 1,326.74m^2 Total Floor Area : 1,106.67m^2

Single Family House		Shinagawa, Tokyo, Japan
2011	**House in Shinagawa**	
		Site Area : 158.81m^2 Total Floor Area : 216.87m^2

Single Family House		Setagaya, Tokyo, Japan
2009	**House in Seijo**	
		Site Area : 301.42m^2 Total Floor Area : 225.14m^2

Single Family House	**Taka House**	Nagoya, Aichi, Japan
2007		
		Site Area : 413.79m²
		Total Floor Area : 333.27m²

Single Family House	**Villa Nakakaruizawa**	Nakakaruizawai, Nagano, Japan
2004		
2006 Selected Work(Design Award), Japan Institute of Architects		Site Area : 1,043.37m² Total Floor Area : 279.85m²

Orphanage	**Aster Orphanage**	Chofu, Tokyo, Japan
2005		
Good Design Award 2012		Site Area : 224.00m² Total Floor Area : 176.22m²

list of works, Jun Watanabe & Associates

1990	Founded, Kroisberg, Berlin Urban Design
1991	N House, Nara Convention Center
1992	Niigata City Cultural Complex
1994	Tohoku Historical Museum
1995	W House, Taichung Civic Center
1996	Kansaikan National Library
	T Company Headquarter Building
1998	Evergreen Laurel Hotel Osaka, House in Kibi
1999	Cow Barn in Appi Resort
2000	Sasebo Ferry Terminal Building
2001	Chubu University Foreign Student Dormitory
	Villa Gamagori
2002	Izumiyamasou Elderly People Nursery Home
2003	Villa Nakakaruizawa
	Osaka Sumiyoshiku Civic Center
2004	Toki Downtown Community Center
	Orphanage in Chofu
	Meguro 2-chome Condominium
2005	House in Nagoya
2006	Michino-eki Tomika
	Condominium in Takatsu
2007	Ogikubo Flat
	House in Hibarigaoka
2008	Hiroo Flat
	House in Seijo
	Civic Center in Kashiwazaki
2009	Ohtaki Town Hall Annex
	Renovation of International Congress Center in Makuhari
2010	Villa Shimazuyama
	House in Yokohama
	Maebashi International College, Student Center
2011	House in Kitamagome
	House in Shinagawa
2012	Kyorin University New Campus
	House in Sugacho
	Doshisha University Chapel

biography, Jun Watanabe

1954	Born in Tokyo
1978	Bachelor of Architecture with High Distinction, Tokyo University
1978-81	Kenzo Tange & URTEC
1983	Master of Architecture, Harvard University
1983-85	I.M.Pei & Partners, New York
1985-90	Maki & Associates
1990-	Jun Watanabe & Associates, Principal

honors and awards

1978	Sotsugyokeikakusho Award, Tokyo University
1981	Fulbright Scholarship
1993	Citation of Honor, American Institute of Architects, Austin Chapter
1995	Design Merit, American Institute of Architects, Austin Chapter
1996	Tenure Professorship, The University of Texas at Austin
1996	Professorship, Chubu University
2001	Selected Work (Design Award) Architectural Institute of Japan
2002	Selected Work (Design Award) Architectural Institute of Japan
2003	Gold Medal, Bienal Miami+Beach Residential Category
2005	Selected Work (Design Award) Japan Institute of Architect
2006	Selected Work (Design Award) Japan Institute of Architect
2010	Selected Work (Design Award) Japan Institute of Architect
2011	Good Design Award
2012	Good Design Award

detailing

Villa Gamagori

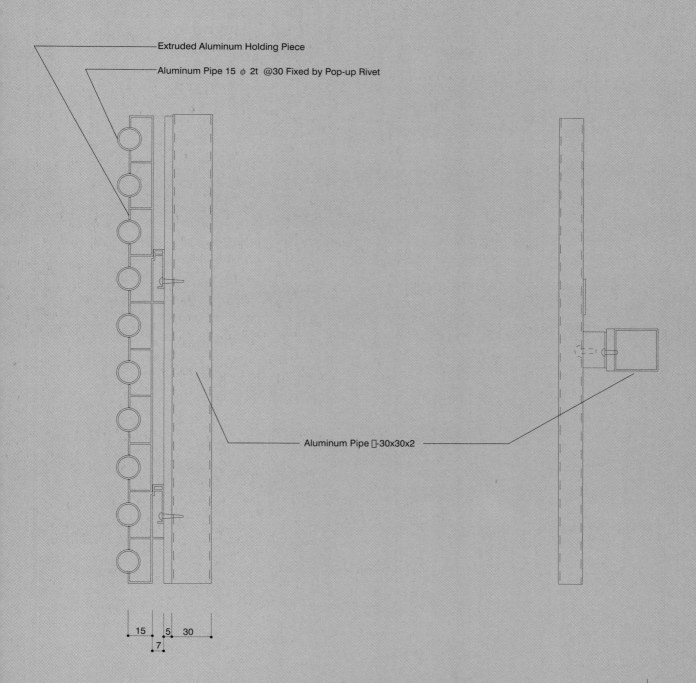

157

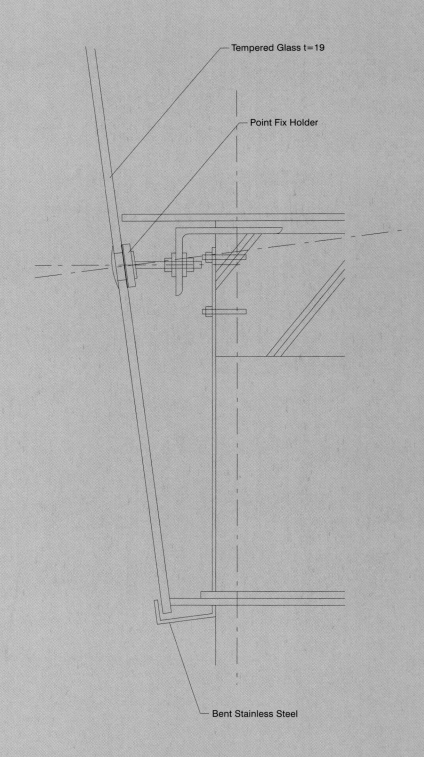

House in Hibarigaoka

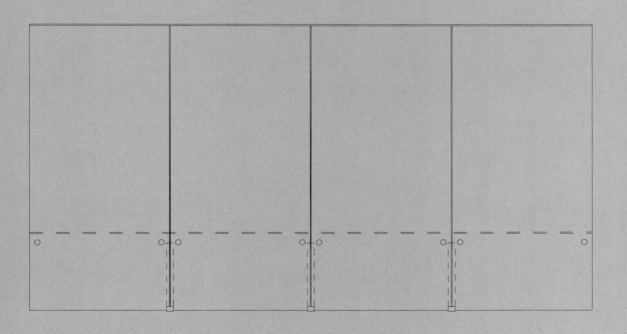

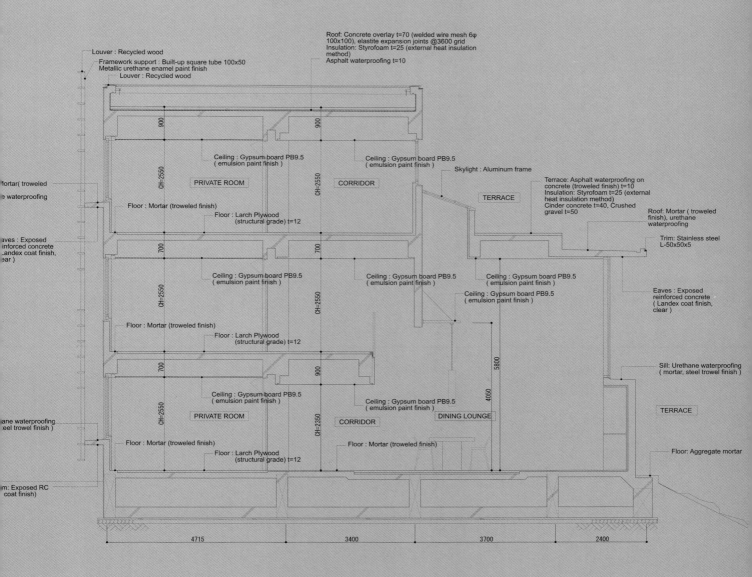

Steel FB9x38 OP
Steel FB9x38 OP
Steel FB9x38 OP
Steel FB9x38 OP
Stainless Steel FB6x38 HL
Larch Plywood t=12
Urethane Lacquer
Structural Plywood Panal t=12

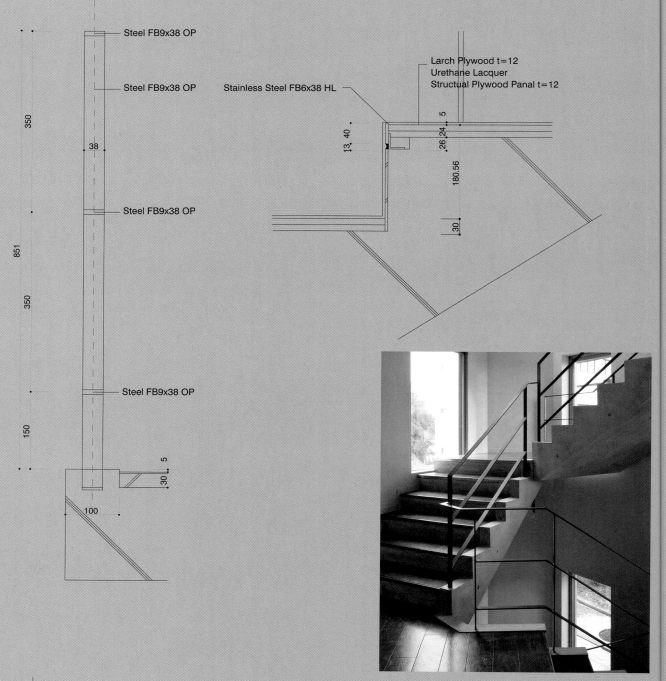

Foreign Student Dormitory, Chubu University

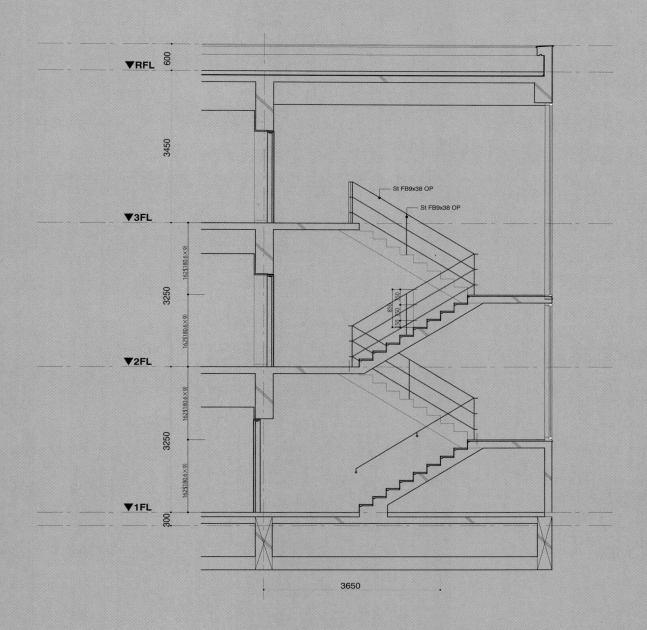

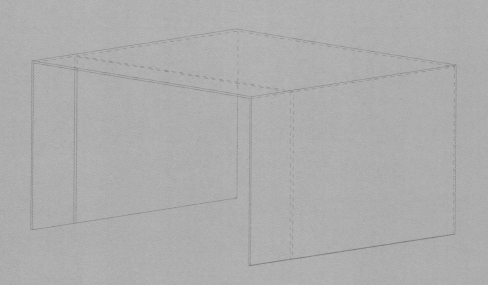

Renovation of the International Congress Hall, Makuhari Messe

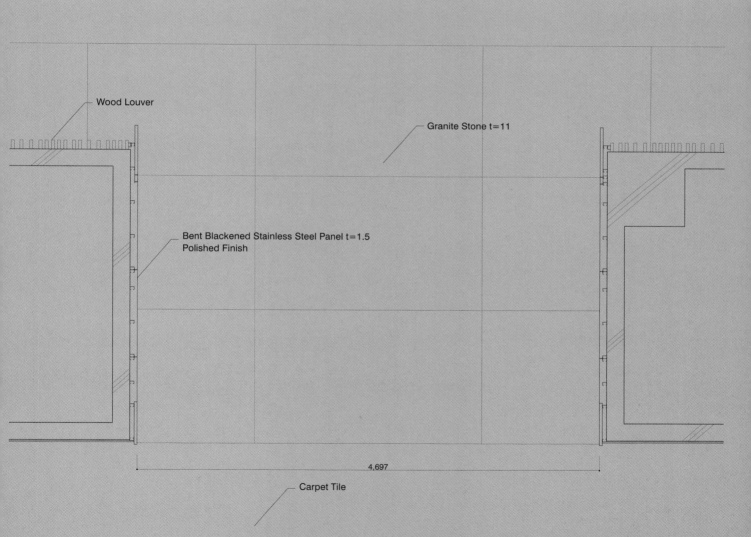

Photo Credits:

Toshiharu Kitajima	pp. 10,17,18,19,22,23,24,25,27,37,38,42,43,45,46,47,48,49,52,53,54,55,58,62,63,65,67,68,69,72,73,75,76,77, 78,79,93,95,97,98,99,102,105,106,107,110,124,128,129,130,131,132,133,136,137,138,139,143,144,145,156, 159,160,162
Makoto Osawa	pp. 30,31,32,33,36
Junji Kojima	pp. 103,108,109,111
Jun Watanabe	pp. 8,21,26,57,92,94,115,116,117,118,119,120,121,125,141,142